The Journey to the Moon

Inside the Race to Be the First

Commercial Spacecraft to Land the Moon

and What's Ahead for Lunar Exploration

Mary E. Boswell

Table of Contents

Introduction

The 21st century has witnessed a remarkable shift in the landscape of space exploration, marked by the rise of private companies venturing beyond Earth's atmosphere. No longer confined to the realm of government agencies, space exploration has become a playground for innovators, entrepreneurs, and visionaries seeking to push the boundaries of human achievement. This introduction sets the stage for delving into the fascinating world of private space exploration, with a particular focus on the groundbreaking Lunar Odyssey mission.

The Dawn of Private Space Exploration

The dawn of private space exploration can be traced back to the early 2000s when a handful of ambitious entrepreneurs began to see the potential for commercial ventures beyond Earth's orbit. Spearheaded by the likes of Elon Musk's SpaceX, Jeff Bezos' Blue Origin, and Richard Branson's Virgin Galactic, these companies set out to revolutionize space travel by leveraging innovation, technology, and entrepreneurial spirit.

One of the defining moments in the rise of private space exploration came in 2004 when SpaceX was founded with the ambitious goal of reducing the cost of space travel and ultimately enabling human colonization of other planets. Through a series of bold initiatives, including the development of the Falcon 1 rocket and the Dragon spacecraft, SpaceX quickly established itself as a frontrunner in the private space industry.

Meanwhile, other companies like Blue Origin and Virgin Galactic were making significant strides in suborbital space tourism, paving the way for a new era of commercial space travel. Blue Origin's New Shepard and Virgin Galactic's SpaceShipTwo promised to offer civilians the opportunity to experience weightlessness and glimpse the curvature of the Earth from the edge of space.

As private companies continued to push the boundaries of space exploration, governments began to recognize the potential for collaboration and partnership. Initiatives like NASA's Commercial Crew Program and Commercial Lunar Payload Services (CLPS) program sought to leverage the expertise and resources of private companies to accelerate progress in space exploration.

Overview of the Lunar Odyssey Mission

Against this backdrop of burgeoning innovation and collaboration, the Lunar Odyssey mission emerges as a pioneering endeavor in the realm of private lunar exploration. Spearheaded by Houston-based company Intuitive Machines, Lunar Odyssey aims to achieve a historic milestone: landing the first commercially built spacecraft on the surface of the Moon.

The mission, named after the legendary Greek hero Odysseus, embodies the spirit of adventure, exploration, and discovery. Scheduled to launch from NASA's Kennedy Space Center in Cape Canaveral, Florida, aboard a SpaceX Falcon 9 rocket, Lunar

Odyssey represents the culmination of years of planning, development, and testing.

The primary objective of the Lunar Odyssey mission is to demonstrate the feasibility and capabilities of commercial lunar landers, paving the way for future missions to the Moon and beyond. Equipped with a mix of commercial cargo and NASA science instruments, the spacecraft aims to touch down near the Moon's south pole, a region of particular interest due to the presence of water ice within craters.

The success of the Lunar Odyssey mission holds profound implications for the future of space exploration. Not only does it mark a significant milestone in the commercialization of space, but it also lays the groundwork for expanded scientific research, resource utilization, and human exploration beyond Earth's orbit.

Chapter 1: The Players

Intuitive Machines: Pioneers of Private Lunar Exploration

Founded in 2013, Intuitive Machines quickly emerged as a leading force in the burgeoning field of private space exploration. Headquartered in Houston, Texas, the company is driven by a mission to revolutionize space technology and expand humanity's presence beyond Earth. At the forefront of Intuitive Machines' ambitious agenda is the development of innovative lunar landers capable of delivering payloads to the surface of the Moon.

One of the company's most notable achievements is its selection as a partner in NASA's Commercial Lunar Payload Services

(CLPS) program. Through this partnership, Intuitive Machines secured a contract to deliver scientific instruments and payloads to the lunar surface, paving the way for commercial lunar exploration.

Central to Intuitive Machines' success is its commitment to innovation and collaboration. The company leverages cutting-edge technology, including advanced propulsion systems, autonomous navigation, and precision landing capabilities, to develop reliable and cost-effective lunar landers. Moreover, Intuitive Machines collaborates closely with government agencies, research institutions, and commercial partners to advance the frontiers of space exploration.

The Lunar Odyssey mission represents a culmination of Intuitive Machines' pioneering efforts in private lunar exploration. Named after the legendary Greek hero Odysseus, the mission embodies the spirit of adventure, discovery, and perseverance. With the launch of the Lunar Odyssey spacecraft, Intuitive Machines aims to make history by becoming the first commercial company to land a spacecraft on the surface of the Moon.

However, the journey to lunar exploration is not without its challenges. Intuitive Machines faces numerous technical, logistical, and regulatory hurdles along the way. From the complexities of rocket propulsion to the intricacies of lunar navigation, the company must navigate a myriad of obstacles to achieve its ambitious goals. Yet, through determination, ingenuity, and sheer grit, Intuitive Machines remain undeterred in its quest to unlock the mysteries of the Moon.

Astrobotic Technology: Challenges and Setbacks

In the competitive landscape of private space exploration, Astrobotic Technology stands out as a pioneering company with a bold vision for the future of lunar exploration. Founded in 2007 by Carnegie Mellon University professor Red Whittaker, Astrobotic aims to make the Moon accessible to commercial and scientific interests alike.

From the outset, Astrobotic faced a daunting challenge: to develop affordable and reliable lunar landers capable of delivering payloads to the lunar surface. With a focus on innovation and efficiency, the company embarked on an ambitious journey to revolutionize space exploration and expand humanity's presence beyond Earth.

One of Astrobotic's most ambitious projects is its participation in NASA's Commercial Lunar Payload Services (CLPS) program. Through this partnership, the company secured a contract to deliver scientific instruments and payloads to the Moon, positioning itself at the forefront of commercial lunar exploration.

However, Astrobotic's journey has been fraught with challenges and setbacks. In 2023, the company suffered a crippling setback when its Peregrine lunar lander experienced a fuel leak shortly after launch, forcing the mission to be aborted. The failure dealt a significant blow to Astrobotic's reputation and raised questions about the viability of its approach to lunar exploration.

Despite the setback, Astrobotic remains committed to its mission of lunar exploration. The company continues to refine its technology, improve its processes, and learn from its mistakes in pursuit of its ambitious goals. With perseverance, resilience, and a steadfast commitment to excellence, Astrobotic is determined to overcome the challenges that lie ahead and make its mark on the future of space exploration.

In conclusion, Intuitive Machines and Astrobotic Technology represent two pioneering companies at the forefront of private lunar exploration. Despite facing numerous challenges and setbacks along the way, both companies remain committed to their respective missions and are poised to play a key role in shaping the future of space exploration.

Chapter 2: Preparing for Launch

From Concept to Reality: Designing Odysseus

The story of Odysseus, the spacecraft tasked with landing on the lunar surface, begins with a bold vision and a team of dedicated engineers and scientists determined to turn that vision into reality. Conceived by Intuitive Machines, Odysseus represents the culmination of years of research, innovation, and ingenuity aimed at pushing the boundaries of space exploration.

The design process for Odysseus is a complex and iterative journey that begins with defining the mission objectives and requirements. Engineers meticulously analyze every aspect of the spacecraft's design, from its propulsion systems and navigation capabilities to its payload capacity and thermal control mechanisms. Drawing on the latest advances in aerospace technology, they strive to create a spacecraft that is not only capable of surviving the harsh conditions of space but also of achieving its ambitious goals.

Central to the design of Odysseus is a focus on reliability, efficiency, and safety. Engineers employ state-of-the-art computer-aided design (CAD) software to model and simulate the spacecraft's performance under various conditions, allowing them to identify potential issues and refine the design before fabrication begins. The result is a sleek and sophisticated spacecraft optimized for performance and reliability in the harsh environment of space.

But designing Odysseus is only half the battle. Turning that design into a physical spacecraft requires precision engineering, rigorous testing, and meticulous attention to detail. Engineers work tirelessly to fabricate, assemble, and integrate the myriad components that make up the spacecraft, from its propulsion systems and avionics to its scientific instruments and landing gear.

As the spacecraft takes shape, engineers conduct a series of comprehensive tests to ensure that every system and subsystem performs as expected. From structural integrity tests to thermal vacuum testing, each test is designed to simulate the rigors of space and identify any potential issues that could jeopardize the mission's success.

Throughout the design and fabrication process, safety is paramount. Engineers implement redundant systems and fail-safe mechanisms to minimize the risk of mission failure and ensure the safety of the spacecraft and its payload. Every component is subjected to rigorous quality control measures to detect and address any defects or anomalies that could compromise performance.

As launch day approaches, the culmination of years of hard work, dedication, and innovation is finally within reach. Odysseus stands ready to embark on its historic journey to the Moon, a testament to the ingenuity, perseverance, and spirit of exploration that define humanity's quest to reach for the stars.

Testing and Safety Measures

The success of any space mission hinges on the thoroughness of its testing and safety measures. For Odysseus, the spacecraft tasked with landing on the lunar surface, ensuring its readiness for the journey to the Moon requires a rigorous regime of testing and safety protocols designed to identify and mitigate any potential risks or issues that could jeopardize the mission's success.

One of the key components of Odysseus' testing regimen is environmental testing, which simulates the harsh conditions of space to evaluate the spacecraft's performance under extreme temperatures, vacuum conditions, and radiation exposure. Thermal vacuum testing, for example, subjects the spacecraft to the temperature extremes and vacuum conditions of space to verify its ability to withstand the rigors of the lunar environment.

In addition to environmental testing, engineers conduct a series of functional tests to verify the performance of the spacecraft's systems and subsystems under normal operating conditions. From propulsion tests and avionics checks to communications tests and payload integration, each test is designed to ensure that every aspect of the spacecraft functions as intended and meets the mission's requirements.

Safety is a top priority throughout the testing process, with engineers implementing stringent safety protocols to protect personnel, equipment, and the environment. Test facilities are equipped with state-of-the-art safety systems and procedures to minimize the risk of accidents and ensure the safe conduct of tests.

As launch day approaches, the culmination of months or even years of testing, Odysseus undergoes a final series of checks and inspections to verify its readiness for flight. Engineers meticulously review every aspect of the spacecraft's design, construction, and testing history to identify any potential issues or concerns that could impact the mission's success.

With the completion of testing and safety measures, Odysseus is poised for launch, ready to embark on its historic journey to the Moon. As it hurtles through the depths of space, carrying the hopes and dreams of countless individuals, the spacecraft stands as a testament to the ingenuity, perseverance, and dedication of the team that brought it to life.

Chapter 3: Countdown to Liftoff

The Initial Launch Delay: Overcoming Challenges

The journey to space is fraught with challenges, and the launch of the Lunar Odyssey mission is no exception. As the countdown to liftoff begins, engineers and technicians at NASA's Kennedy Space Center in Cape Canaveral, Florida, face a series of unexpected setbacks that threaten to derail the mission before it even begins.

One of the most significant challenges comes in the form of a glitch with the rocket's methane fuel, forcing SpaceX to postpone the launch initially scheduled for Wednesday. The delay sends ripples of uncertainty and frustration through the mission control center as teams scramble to identify the cause of the issue and develop a plan to address it.

Despite the setback, the teams remain undeterred, drawing on their expertise, experience, and determination to overcome the challenges they face. Engineers work around the clock to troubleshoot the issue, conducting a thorough analysis of the rocket's systems and components to pinpoint the source of the problem and implement a solution.

As the hours tick by, tensions mount as the fate of the mission hangs in the balance. Yet, amidst the uncertainty and adversity, the teams remain focused on their ultimate goal: to successfully launch the Lunar Odyssey spacecraft and pave the way for a new era of lunar exploration.

After hours of intensive effort and collaboration, the teams finally identify and resolve the issue with the rocket's methane fuel, clearing the way for the mission to proceed. The initial launch delay serves as a testament to the resilience and ingenuity of the teams involved, demonstrating their ability to overcome obstacles and adapt to unforeseen challenges in pursuit of their goals.

Final Preparations at Kennedy Space Center

With the initial launch delay overcome, attention turns to the final preparations for the launch of the Lunar Odyssey mission at NASA's Kennedy Space Center. As the countdown clock ticks inexorably towards zero, engineers, technicians, and support personnel work feverishly to ensure that everything is ready for liftoff.

At the launch pad, technicians conduct a series of final checks and inspections to verify the readiness of the Falcon 9 rocket and the Lunar Odyssey spacecraft. Every system and subsystem is scrutinized for any signs of anomalies or issues that could compromise the mission's success.

Meanwhile, mission control teams monitor weather conditions and track the trajectory of the rocket to ensure optimal launch conditions. A favorable weather forecast is essential for a successful launch, and teams closely monitor atmospheric conditions to determine the ideal window for liftoff.

Inside the mission control center, engineers and flight controllers conduct a series of simulations and rehearsals to prepare for every possible scenario that could arise during the launch and ascent phase. From contingency plans for technical malfunctions to procedures for aborting the mission in the event of an emergency, every detail is carefully considered and meticulously planned.

As the final minutes tick away, anticipation mounts as the launch pad comes alive with activity. Engineers perform a final sweep of the area, ensuring that all personnel are clear of the danger zone and that all systems are ready for liftoff.

Then, at precisely T-0, the Falcon 9 rocket roars to life, its engines igniting with a deafening roar as it lifts off from the launch pad, carrying the

Lunar Odyssey spacecraft towards its destination in the depths of space.

As the rocket disappears into the sky, a collective sigh of relief sweeps through the mission control center as teams watch with bated breath, knowing that their hard work, dedication, and perseverance have paid off. The countdown to liftoff may be over, but the journey to the Moon has only just begun.

Chapter 4: Liftoff and Journey to Orbit

A Night to Remember: Odysseus Takes Flight

As the clock ticks down to zero and anticipation fills the air, the moment of liftoff arrives, signaling the beginning of an extraordinary journey into the depths of space. Against the backdrop of the night sky, illuminated by the glow of the Falcon 9 rocket, Odysseus takes flight, its engines roaring to life with a deafening roar that reverberates across the launch pad.

For the teams at mission control, the moment is both exhilarating and nerve-wracking as they monitor every aspect of the launch with precision and focus. Engineers and flight controllers track the trajectory of the rocket, ensuring that it follows the planned flight path and reaches the desired orbit.

As the Falcon 9 rocket ascends into the sky, leaving a trail of fire and smoke in its wake, spectators on the ground watch in awe as Odysseus disappears into the darkness, embarking on a journey that will take it hundreds of thousands of miles from Earth.

For the engineers and technicians who have dedicated countless hours to the development and preparation of Odysseus, the moment of liftoff is a culmination of years of hard work, determination, and perseverance. It is a testament to their skill, expertise, and unwavering commitment to pushing the boundaries of space exploration.

As Odysseus soars higher and higher into the sky, it carries with it the hopes and dreams of countless individuals who have dared to imagine a future where humanity explores the cosmos and unlocks the secrets of the universe. It is a night to remember, a moment that will be etched in the annals of history as humanity takes another bold step towards the stars.

Riding the Falcon 9: Collaboration with SpaceX

Central to the success of the Lunar Odyssey mission is the collaborative partnership between Intuitive Machines and SpaceX, two pioneering companies at the forefront of space exploration. Together, they have worked tirelessly to bring Odysseus to life and propel it towards its destination in the depths of space.

At the heart of this collaboration is the Falcon 9 rocket, SpaceX's workhorse launch vehicle that has revolutionized space transportation with its reliability, efficiency, and cost-effectiveness. Designed to deliver payloads to low Earth orbit, geostationary transfer orbit, and beyond, the Falcon 9 rocket is ideally suited for launching Odysseus on its journey to the Moon.

Throughout the development and preparation of the Lunar Odyssey mission, Intuitive Machines and SpaceX have worked closely together to ensure that every aspect of the launch goes smoothly. From coordinating launch schedules and conducting pre-flight checks to integrating the spacecraft with the rocket and monitoring the launch in real-time, the collaboration between the two companies has been seamless and efficient.

As Odysseus rides atop the Falcon 9 rocket, it is a testament to the power of collaboration and partnership in the pursuit of ambitious goals. Together, Intuitive Machines and SpaceX have overcome countless challenges and obstacles to bring the Lunar Odyssey mission to fruition, demonstrating the potential of private companies to drive innovation and progress in space exploration.

As the Falcon 9 rocket carries Odysseus towards orbit, it symbolizes the collective efforts of engineers, scientists, and visionaries who have dared to dream of a future where humanity explores the cosmos and reaches for the stars. It is a moment of triumph, a celebration of ingenuity, and a testament to the power of collaboration to achieve the impossible.

Chapter 5: Onward to the Moon

In Transit: Odysseus' Journey through Space

With the initial launch phase complete, Odysseus begins its journey through the vast expanse of space, hurtling towards its destination with precision and purpose. As it travels through the void, the spacecraft encounters a series of challenges and obstacles, from micro-meteoroids and radiation to the harsh conditions of space itself.

To navigate safely through space, Odysseus relies on a sophisticated array of sensors, cameras, and navigation systems that enable it to precisely calculate its trajectory and adjust its course as needed. Autonomous control systems ensure that the spacecraft remains on track, constantly monitoring its position and velocity relative to its target.

As Odysseus traverses the vast distances between Earth and the Moon, it passes through the Van Allen radiation belts, a region of intense radiation surrounding our planet. Shielded by layers of protective insulation and shielding, the spacecraft withstands the onslaught of radiation as it continues its journey towards its destination.

Despite the challenges of space travel, Odysseus presses onward, fueled by the spirit of exploration and the determination to reach its destination. Along the way, it gathers valuable data and scientific observations that will help scientists better understand the complexities of space and pave the way for future missions to the Moon and beyond.

As the days turn into weeks and the weeks turn into months, Odysseus draws ever closer to the Moon, its destination growing larger in the viewport with each passing moment. With anticipation building and excitement mounting, the spacecraft prepares for its final approach to the lunar surface, where it will fulfill its mission objectives and make history as the first commercially built spacecraft to land on the Moon.

Mission Objectives and Cargo

At the heart of the Lunar Odyssey mission are a set of ambitious objectives aimed at advancing our understanding of the Moon and unlocking its potential for scientific discovery and exploration. From studying the lunar surface to conducting experiments and deploying instruments, Odysseus carries out a variety of tasks designed to achieve these objectives and pave the way for future missions to the Moon.

One of the primary objectives of the Lunar Odyssey mission is to study the composition and geology of the lunar surface, shedding light on its formation and evolution over billions of years. Equipped with a suite of scientific instruments, including cameras, spectrometers, and seismometers, Odysseus gathers data and images that will help scientists better understand the Moon's geological history and identify potential sites for future exploration.

In addition to studying the lunar surface, Odysseus also carries out experiments aimed at testing new technologies and techniques for future lunar missions. From testing the performance of new propulsion systems to evaluating the effectiveness of radiation shielding materials, these experiments provide valuable insights that will inform the design and development of future spacecraft and habitats for lunar exploration.

Furthermore, Odysseus serves as a cargo transport vehicle, delivering a variety of payloads to the lunar surface as part of NASA's Commercial Lunar Payload Services (CLPS) program. These payloads include scientific instruments, technology demonstrations, and commercial cargo, all of which play a crucial role in advancing our understanding of the Moon and paving the way for future human exploration.

As Odysseus approaches the Moon's surface, anticipation reaches a fever pitch as mission control teams closely monitor its progress and prepare for the historic moment of touchdown. With its cargo safely secured and its mission objectives within reach, Odysseus stands poised to make history as it touches down on the lunar surface, marking a milestone in the annals of space exploration and paving the way for a new era of lunar exploration and discovery.

Chapter 6:Touchdown at the South Pole

The Final Approach: Navigating Lunar Terrain

As Odysseus nears the lunar south pole, it enters a realm of rugged terrain and treacherous obstacles, where every decision and maneuver is critical to the success of the mission. The final approach to the landing site is fraught with challenges, from towering mountains and deep craters to jagged rocks and uneven surfaces.

To navigate safely through this hostile environment, Odysseus relies on a combination of autonomous navigation systems and real-time data feedback from mission control. Advanced sensors and cameras scan the lunar surface, mapping out potential hazards and identifying suitable landing sites with pinpoint accuracy.

As Odysseus descends towards the lunar surface, its onboard computers analyze the terrain in real-time, adjusting its trajectory and descent profile to avoid obstacles and ensure a safe landing. Engineers and flight controllers monitor the spacecraft's progress closely, ready to intervene at a moment's notice if any issues arise.

Despite the challenges of navigating lunar terrain, Odysseus presses onward with determination and precision, drawing ever closer to its destination with each passing moment. The final approach is a testament to the ingenuity and expertise of the mission's engineers and scientists, who have meticulously planned and prepared for every possible scenario.

As the spacecraft descends towards the lunar surface, tension mounts in mission control as teams watch anxiously, knowing that the success of the entire mission hinges on the outcome of the final approach. Every second feels like an eternity as Odysseus inches closer to touchdown, with the fate of the mission hanging in the balance.

Landing on the Lunar Surface: Success or Setback?

As Odysseus makes its final descent towards the lunar surface, the tension in mission control reaches a fever pitch as teams anxiously await the outcome of the landing. Every heartbeat feels like a drum roll as engineers and flight controllers hold their breath, hoping for a successful touchdown that will mark a historic milestone in space exploration.

With bated breath, the moment of truth arrives as Odysseus' landing legs make contact with the lunar surface, absorbing the impact of touchdown with remarkable precision. Mission control erupts into cheers and applause as telemetry data confirms that the spacecraft has safely landed on the Moon, marking a resounding success for the mission.

For the engineers, scientists, and visionaries who have dedicated years of their lives to making this moment possible, the feeling of triumph is overwhelming as they celebrate the culmination of their hard work, dedication, and perseverance. It is a moment of joy and pride as Odysseus stands proudly on the lunar surface, a testament to human ingenuity and the spirit of exploration.

However, landing on the lunar surface is not without its risks, and for some missions, the outcome is not always as hoped. Despite meticulous planning and preparation, unforeseen challenges or technical issues can sometimes lead to a less-than-ideal landing or, in some cases, mission failure.

In the event of a setback, engineers and scientists must regroup, analyze what went wrong, and learn from their mistakes to ensure that future missions are more successful. Failure is an inevitable part of the journey of exploration, but it is also an opportunity to grow, adapt, and improve for the future.

Fortunately, for the Lunar Odyssey mission, the landing is a resounding success, marking a historic milestone in space exploration and paving the way for future missions to the Moon and beyond. As Odysseus begins its mission of scientific discovery and exploration, it carries with it the hopes and dreams of countless individuals who dare to reach for the stars and unlock the secrets of the cosmos.

Chapter 7: The Significance of Success

Implications for Lunar Exploration

The successful landing of Odysseus at the lunar south pole opens up a wealth of opportunities for scientific discovery and exploration on the Moon. Equipped with a suite of scientific instruments and payloads, the spacecraft is poised to study the composition, geology, and environment of the lunar surface in unprecedented detail, shedding light on the mysteries of our celestial neighbor.

One of the primary objectives of the Lunar Odyssey mission is to study the presence of water ice at the lunar south pole, a valuable resource that could potentially support future human exploration and habitation of the Moon.

By analyzing samples of lunar regolith and conducting experiments on the surface, Odysseus aims to confirm the presence of water ice and assess its abundance and accessibility, laying the groundwork for future missions to extract and utilize this valuable resource.

Furthermore, the success of the Lunar Odyssey mission paves the way for future human exploration of the Moon, providing valuable data and insights that will inform the design and planning of crewed missions to the lunar surface. By studying the lunar environment and conducting experiments on the surface, Odysseus helps scientists better understand the challenges and opportunities of living and working on the Moon, from radiation exposure and dust mitigation to resource utilization and habitat construction.

Moreover, the successful landing of Odysseus at the lunar south pole reaffirms the importance of international collaboration in space exploration. By partnering with private companies like Intuitive Machines and leveraging their expertise and resources, NASA and other space agencies are able to achieve ambitious goals and milestones that would be impossible to accomplish alone. The Lunar Odyssey mission serves as a shining example of the power of collaboration and cooperation in advancing the frontiers of space exploration.

Commercial Opportunities and Future Endeavors

In addition to its scientific significance, the successful landing of Odysseus at the lunar south pole unlocks a wealth of commercial opportunities and future endeavors in space. As private companies like Intuitive Machines and SpaceX continue to demonstrate their capabilities in space exploration, the door opens for a new era of commercialization and economic activity in low Earth orbit and beyond.

One of the most promising opportunities for commercial exploitation is the extraction and utilization of resources on the Moon, including water ice, rare minerals, and precious metals. By establishing mining operations and processing facilities on the lunar surface, private companies can tap into these valuable resources and potentially generate substantial revenue streams while supporting future human exploration and settlement of the Moon.

Furthermore, the successful landing of Odysseus at the lunar south pole opens up opportunities for tourism and space travel, with private companies like SpaceX and Blue Origin already offering commercial flights to space for paying customers. As the cost of space travel continues to decline and access to space becomes more accessible, the possibility of lunar tourism becomes increasingly feasible, with tourists and adventurers flocking to the Moon to experience the thrill of space exploration firsthand.

Moreover, the success of the Lunar Odyssey mission serves as a catalyst for future commercial endeavors in space, from satellite deployment and telecommunications to in-orbit servicing and asteroid mining. As private companies continue to innovate and invest in space exploration, the possibilities for commercialization and economic growth in space are virtually limitless, paving the way for a vibrant and sustainable space economy in the decades to come.

In conclusion, the successful landing of Odysseus at the lunar south pole represents a watershed moment in the history of space exploration, with profound implications for lunar exploration and commercial opportunities in space. By unlocking the mysteries of the Moon and paving the way for future endeavors in space, the Lunar Odyssey mission sets the stage for a new era of discovery, innovation, and economic growth in the cosmos.

Chapter 8: Challenges and Lessons Learned

Reflecting on the Journey: Triumphs and Tribulations

The journey to the Moon is one of triumphs and tribulations, marked by moments of exhilaration and moments of heartbreak. For the teams behind the Lunar Odyssey mission, the journey has been a rollercoaster ride of emotions, with highs and lows that have tested their resolve and determination.

One of the greatest triumphs of the Lunar Odyssey mission is the successful landing of Odysseus at the lunar south pole, marking a historic milestone in space exploration and paving the way for future missions to the Moon. It is a moment of joy and pride for the teams involved, a validation of their hard work, dedication, and perseverance in the face of seemingly insurmountable challenges.

However, the journey has not been without its tribulations. Along the way, the teams have encountered a myriad of technical issues, logistical challenges, and unforeseen obstacles that threatened to derail the mission. From launch delays and propulsion problems to navigation errors and communication glitches, each setback tested the teams' resolve and forced them to adapt and innovate in real-time.

Yet, despite the challenges, the teams remained undeterred in their pursuit of the mission's goals. They drew on their collective expertise, creativity, and resilience to overcome each obstacle and forge ahead towards success. It is a testament to the spirit of exploration and the indomitable human spirit that drives us to push the boundaries of what is possible and reach for the stars.

How Failures Shape Future Successes

In the world of space exploration, failure is not only inevitable but also invaluable. Each setback and failure serves as a learning opportunity, providing valuable insights and lessons that shape future successes and innovations. For the teams behind the Lunar Odyssey mission, failures have been a catalyst for growth, driving them to improve and innovate in pursuit of their goals.

One of the most important lessons learned from the challenges encountered during the Lunar Odyssey mission is the importance of resilience and adaptability in the face of adversity. The teams quickly learned to embrace failure as a natural part of the journey and to use it as an opportunity for growth and improvement. They developed contingency plans and backup systems to mitigate the impact of potential failures, ensuring that the mission could continue even in the face of setbacks.

Moreover, failures have also led to important technological innovations and advancements that have shaped the future of space exploration. From improvements in propulsion systems and navigation techniques to advancements in materials science and robotics, each failure has spurred the development of new technologies and techniques that have made space exploration safer, more efficient, and more reliable.

Furthermore, failures have also fostered a culture of collaboration and cooperation in the space industry, with companies and organizations coming together to share knowledge, resources, and expertise in pursuit of common goals. By working together, sharing lessons learned, and pooling resources, the space industry has been able to overcome challenges and achieve milestones that would be impossible to accomplish alone.

In conclusion, the challenges and failures encountered during the Lunar Odyssey mission have been valuable learning experiences that have shaped the future of space exploration. By embracing failure as a natural part of the journey and using it as an opportunity for growth and improvement, the teams behind the mission have demonstrated the resilience, ingenuity, and determination that define humanity's quest to explore the cosmos. As we reflect on the journey, we are reminded that it is not the failures themselves that define us, but how we respond to them and the lessons we learn along the way that ultimately shape our success.

Chapter 9: Looking Ahead

The Next Frontier: NASA's Artemis Program

NASA's Artemis Program represents a bold step forward in space exploration, with the goal of returning humans to the Moon and laying the groundwork for future missions to Mars. Named after the Greek goddess of the Moon and sister of Apollo, the Artemis Program aims to build upon the successes of the Apollo missions while leveraging new technologies and capabilities to enable sustainable lunar exploration.

Central to the Artemis Program is the Artemis mission, which aims to land the first woman and the next man on the lunar surface by 2024. This historic milestone will not only mark a significant achievement in space exploration but will also serve as a stepping stone towards future missions to Mars and beyond.

The Artemis mission will utilize the Space Launch System (SLS), NASA's next-generation heavy-lift rocket, to launch the Orion spacecraft and its crew into space. From there, the crew will embark on a journey to lunar orbit, where they will rendezvous with the Lunar Gateway, a space station that will serve as a staging point for missions to the lunar surface.

Once at the Gateway, astronauts will transfer to a lunar lander, which will carry them to the surface of the Moon. There, they will conduct scientific research, explore the lunar terrain, and test new technologies and systems in preparation for future missions to Mars.

In addition to crewed missions, the Artemis Program also includes a series of robotic missions to the Moon, aimed at conducting scientific research and exploring potential landing sites for future human missions. These missions will provide valuable data and insights into the lunar environment and help inform the design and planning of future exploration activities.

Furthermore, the Artemis Program is a collaborative effort, with international partners, commercial companies, and other space agencies contributing expertise, resources, and capabilities to achieve common goals and objectives. By working together, the Artemis Program aims to maximize the impact and success of lunar exploration and pave the way for a sustainable future in space.

Opportunities and Obstacles on the Path to Lunar Colonization

As humanity sets its sights on returning to the Moon and establishing a sustainable presence there, a world of opportunities and obstacles awaits on the path to lunar colonization.

One of the most promising opportunities for lunar colonization is the abundance of resources available on the Moon, including water ice, rare minerals, and precious metals. These resources could potentially be mined and utilized to support human habitation and exploration, providing essential supplies and materials for sustaining life and conducting research on the lunar surface.

Water ice, in particular, is of great interest to scientists and engineers, as it can be used to produce drinking water, oxygen for breathing, and hydrogen for fuel. By extracting and utilizing water ice from the lunar poles, astronauts could establish self-sustaining habitats and refueling stations, enabling extended stays on the Moon and facilitating future missions to Mars and beyond.

However, the path to lunar colonization is not without its challenges. One of the greatest obstacles is the harsh and unforgiving environment of space, where radiation exposure, microgravity, and extreme temperatures pose significant risks to human health and safety. Protecting astronauts from these hazards will require innovative technologies and engineering solutions, as well as careful planning and preparation for long-duration missions in deep space.

Furthermore, the cost and complexity of lunar exploration and colonization present additional challenges that must be overcome. From the development of new spacecraft and launch vehicles to the construction of habitats and infrastructure on the lunar surface, the logistical challenges of establishing a sustainable presence on the Moon are immense and require coordinated efforts from governments, private companies, and international partners.

Despite these challenges, the opportunities presented by lunar colonization are too great to ignore, offering the promise of scientific discovery, economic growth, and a sustainable future for humanity in space. By working together, pooling resources, and leveraging advancements in technology and innovation, we can overcome the obstacles that stand in our way and embark on a new era of exploration and discovery on the Moon and beyond.

In conclusion, as we look ahead to the future of lunar exploration and colonization, we are filled with both excitement and anticipation for the opportunities that lie ahead. With the Artemis Program leading the way, humanity stands poised to return to the Moon and establish a sustainable presence on its surface, unlocking the mysteries of the cosmos and paving the way for a bright and prosperous future in space.

Conclusion

The Legacy of Lunar Odyssey

The legacy of the Lunar Odyssey mission is one of innovation, collaboration, and discovery. From its inception as an ambitious vision to its successful culmination as the first commercially built spacecraft to land on the Moon, the mission has left an indelible mark on the history of space exploration.

Central to the legacy of the Lunar Odyssey mission is the spirit of innovation and ingenuity that drove its success. From the development of cutting-edge technologies and systems to the execution of complex maneuvers and operations, the mission pushed the boundaries of what was thought possible and demonstrated the power of human creativity and determination in the face of adversity.

Moreover, the mission's success was made possible by the collaborative efforts of a diverse team of engineers, scientists, technicians, and visionaries from around the world. By working together, sharing knowledge and resources, and overcoming obstacles as a united front, the team behind the Lunar Odyssey mission exemplified the spirit of cooperation and collaboration that defines humanity's quest to explore the cosmos.

Furthermore, the legacy of the Lunar Odyssey mission extends beyond its scientific and technological achievements to inspire future generations of explorers, scientists, and engineers. By demonstrating the potential of private companies to drive innovation and progress in space exploration, the mission has paved the way for a new era of commercialization and economic activity in low Earth orbit and beyond.

The Ever-Expanding Frontier of Space Exploration

As we look to the future, the frontier of space exploration continues to expand, offering endless opportunities for discovery, innovation, and adventure. From the Moon to Mars and beyond, humanity's quest to explore the cosmos knows no bounds, driven by a relentless curiosity and a desire to unlock the mysteries of the universe.

Central to the future of space exploration is the Artemis Program, NASA's ambitious initiative to return humans to the Moon and establish a sustainable presence there. With plans to land the first woman and the next man on the lunar surface by 2024, the Artemis Program represents a bold step forward in humanity's quest to explore the cosmos and lay the groundwork for future missions to Mars and beyond.

Moreover, the Artemis Program is just the beginning of humanity's journey into the cosmos. As technology advances and our understanding of the universe grows, new opportunities for exploration and discovery will continue to emerge, from exploring the icy moons of Jupiter and Saturn to venturing beyond the confines of our solar system in search of other habitable worlds.

Furthermore, the future of space exploration holds promise for commercial opportunities and economic growth, with private companies and entrepreneurs playing an increasingly prominent role in driving innovation and progress in space. From launching satellites and conducting research in low Earth orbit to mining asteroids and establishing colonies on Mars, the possibilities for commercialization and economic activity in space are virtually limitless.

In conclusion, as we stand on the threshold of a new era of space exploration, we are filled with both excitement and anticipation for the opportunities that lie ahead. From the legacy of the Lunar Odyssey mission to the promise of the Artemis Program and beyond, the quest to explore the cosmos continues to inspire and captivate humanity's imagination, driving us ever forward towards new frontiers of discovery and adventure.